# IS MAJORITY OF MANKIND

# SUN WORSHIPPER?

## A SHOCKING REALITY OR MYTH?

ALHAMDULILLAH E RAB AL A'ALAMEEN (THANKS TO ALLAH THE ALMIGHTY), I HAVE PUT IN MY BEST AND SINCERE EFFORT TO GET TO THE TRUTH ON THIS SUBJECT.

IT IS AN HUMBLE EFFORT TO OPEN THE DOOR OF KNOWLEDGE FOR THOSE WHO HAVE THE QUEST FOR STRIVING FURTHER TOWARD CONNECTING THE DOTS AND SEEKING TRUTH!

AND I WILL ALWAYS SAY "ALLAH KNOWS BEST!"

I dedicate this book to all "Truth Seekers", who have the quest for seeking knowledge (Truth in every field of knowledge) and intend to follow and propagate the Truth!

I am indebted and grateful to Eric Dubay, who is the source of knowledge and motivation for me and enabled me to put up this humble effort!

May Allah SWT accept my efforts for enlightening and safeguarding the humanity in the Current Age of Deception i.e. Fitan (End Times) –

Ameen Ya Rab Al A'alameen!

# Contents

## PURPOSE

→ The primary purpose of this humble presentation is to **make the humanity alert of the serious and critical fact** that "**unconsciously and unknowingly" they (majority) are following the ideologies / concepts, so-called in the name of "<u>Science or Scientific Facts</u>,"** which is actually <u>leading them to Sun Worshipping / Idolatry!</u> A believer very well knows that such a consequence will lead him / her "no other than but Hell fire on the Judgment Day", **<u>if he / she is unable to make him / her self-aware and repent in the life here-in!!</u>** On the other hand, if a person is not a believer but at least has the **"<u>quest for knowledge i.e., Truth, then he / she will be able to "uncover the deception and safeguard him /her and the near and dear ones"!</u>**

→ **<u>I intend to "bust the dangerous agenda"</u>** being <u>planned not few decades or centuries but at least more than two thousand years ago!!!</u>

→ I can understand that **<u>this is a very tall order and a serious claim</u>,** as there are many **who claim themselves to be the believers in One God (Deity) may be offended**, but <u>facts speak louder than</u>

mere words. Please **stay till the end to "dig into the rabbit hole"** and find out **whether I am right or wrong** and If "**ONE PERSON**" **is able to understand and be able to return to the "PATH OF TRUTH", then my humble effort will serve the purpose in shaa Allah (God Willing)!**

**DEFINITIONS** - Please learn these definitions, as they will be extensively use ahead.......

→ **PSEUDOSCIENCE:** A system of theories, assumptions, and methods erroneously regarded as scientific. [This is presented to be science or scientific facts but actually NOT. **This is merely to deceive the masses**]

→ **GEOCENTRIC:** A cosmological model in which the **Earth is considered to be the center of the solar system.** The Moon, the planets, the Sun, and the stars all rotate around the Earth (which stays still), with uniform circular motion.

→ **HELIOCENTRIC:** A cosmological model in which the **Sun is assumed to lie at or near a central point** (e.g., of the solar system or of the universe) while the Earth and other bodies revolve around it.

→ **GATEKEEPERS:** Gatekeepers are those people who restrict complete or partial "**knowledge / truth**" from being disseminated to the common people for their ulterior motives (hidden agenda). [In other words, they disseminate information which **only suits their agenda**].

**"RA" IS THE SUN god in paganism!**

**MOVEMENT TOWARDS SUN WORSHIPPING, SINCE THOUSANDS OF YEARS AGO!!**

**EGYPTIANS WORSHIPPING "RA god"!**

## ABOUT "FREEMASONS" AND THEIR SUN WORSHIPPING BELIEF!

The **Masons' esoteric religion,** the very basis of their symbols and rituals, is Sun-worship. From their first day in the lodge, Masonic initiates

9

learn that Freemasonry is all about **light, enlightenment, illumination** (hence, the "Illuminati) and <u>hence worship of the Sun as the giver of light.</u> **Masonic halls are all <u>purposely constructed </u>to correspond with the motions of the Sun. They are always situated intentionally facing <u>East towards the Sun, with the "Worshipful Master" sitting in the far East on a throne</u> engraved with a picture of the Sun.**

The high festival of the Masons is on Christian's "St. John's Day," or the **24th of June, otherwise known as "midsummer day," when the Sun arrives at its annual highest elevation, the summer solstice. Regarding the Masonic "Rite of Circumambulation," 33rd degree Freemasonic historian Albert Mackey says, "*In Freemasonry people always walked three times round the altar while singing a sacred hymn. In making this procession, <u>great care was taken to move an imitation of the course of the Sun.</u> <u>This Rite of Circumambulation undoubtedly refers to the doctrine of sun-worship.</u>*"**

**The scope of this presentation is to <u>highlight and raise the awareness</u> amongst the respected viewers about the Freemasons and their core belief of paganism / sun worshipping. However, the viewer may do his / her independent research for further exploration of this subject. As**

10

you proceed ahead, you will find **various concepts / philosophies and so-called famous people associated** with this "secret and sun worshipping society", **shockingly responsible for misleading mankind from the TRUE PATH in core areas of human society!**

In this Book, I will attempt to dig the rabbit hole <u>only</u> in the field of "Core Science / Scientific Concepts" challenging the authentic beliefs (since time and memorial) and **attempt to answer a million-dollar question** that why these Freemasons (Occultists / Paganist / Sun / Ancient gods Worshippers) were heavily involved specifically in this field?

**PYTHAGORAS 500 B.C. - THIS FIRST RECOGNIZED FREEMASON AND SUN WORSHIPPER!**

**WATCH THE SYMBOLS CAREFULLY!!**

## FREEMASONS' LEGACY OF CRAFTING "HELIOCENTRIC MODEL" BY DESIGN

Please See The Bigger Picture And Don't Get Deceived By The So-Called Big Names Amongst The Scientific Community!!!

[Please Watch Inside The Red Circles, Any Similarity?]

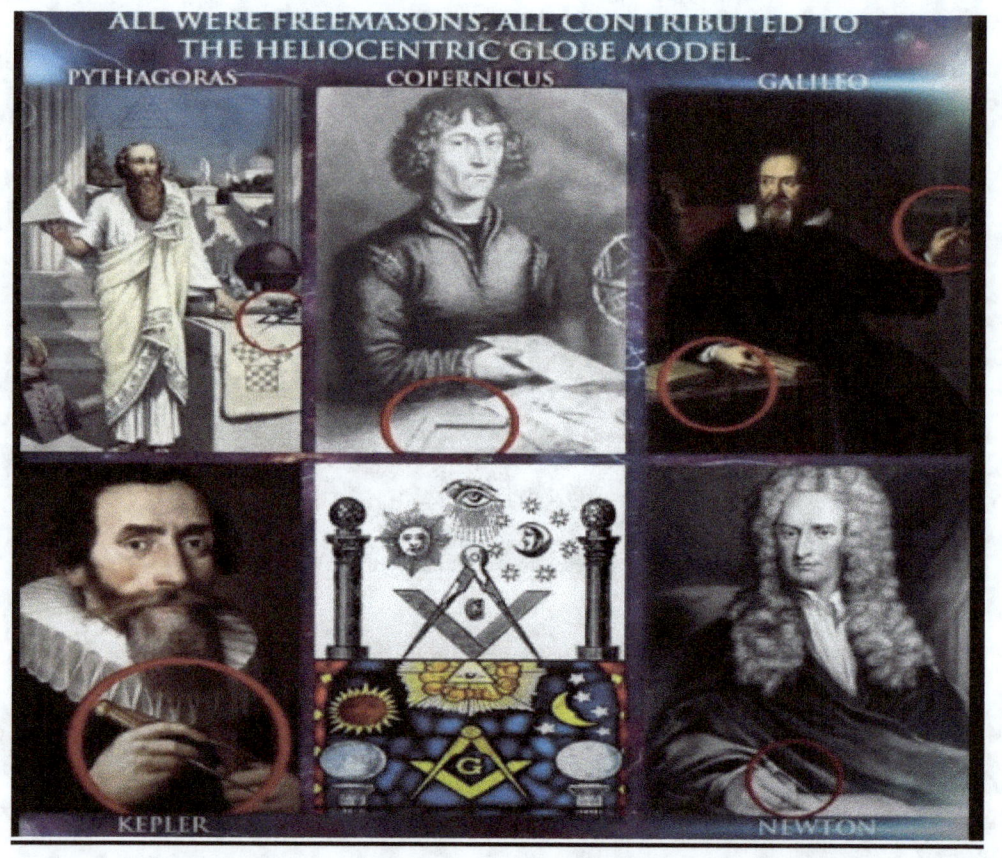

**Please check the square and the compass in all pictures. these symbols are linked to the sacred geometry of the freemasons and all of these men were Freemasons! starting from Pythagoras till Newton. All five of these men were the architect of "Heliocentric - Sun Centered Model". These were "actually sun worshippers" because the sun is the Center of Freemasonry!!**

**So, please don't just take them as mere "Science People or Philosophers" meant for the enlightenment of humanity by**

**discovering some laws or imparting pseudo knowledge**, rather they have an established connection with the secret society (freemason), which has a clear and dictated agenda to deceive the mankind to achieve their ulterior objectives in a gradual, scientific and coordinated way:

1. First to make people believe that they are a useless speck of dust in the universe with no purpose in life.

2. Second, to ultimately become their masters by controlling them and the resources of the world (period)!

## "ISAAC NEWTON's" CONNECTION WITH SUN WORSHIPPING / PAGANISM

# Isaac Newton And Hermes Trismegistus Revisited

The writings of Hermes Trismegistus was revered by Isaac Newton. It was Newton who developed the theory of gravity. Newton theories were heavily influenced by the writings of Hermes Trismegistus. Among Newton's notes were found the following quote in which he translated the writing of Hermes Trismegistus. **In these notes, Newton wrote that the Sun is the Father of creation, the moon is the mother, and the earth is the nurse. It was Hermes Trismegistus spinning ball (spherical earth) that obvio**usly led to the need for gravity. Without gravity, it would be impossible to explain a spherical earth in which people could walk on a spinning ball without falling off.

## JESUITS (SOCIETY OF JESUS) CONNECTION WITH "HELIOCENTRISM"

**Jesuits (Society of Jesus) had been involved in A Plan toward moving from "Geocentrism" (Truth) To "Heliocentrism - Sun Worshipping"!!**

This Is the **Darkest Side of Christianity**, who appeared to be Christians but heavily involved in black magic and satanism and **nothing to do with the true teachings of Christianity!** This **powerful but hidden group** have **penetrated in all walks of societies** and **even control the Catholic Pope and Elites Of The World!!** Check their Logo on the **Left- Illuminating Sun.** It may not be incorrect to say that with **Their (Jesults)** "Secret Infiltration" In the Church, Catholicism Gradually Moved Toward Satanism / Luciferins / Idolatry and unfortunately masses have "**No Or Very Little Clue**" in this regard! Hence, **movement toward paganism in a secretive and gradual way!!!**

**ROMAN CATHOLICISM TURNED TO PAGANISM!**

THEY ACTUALLY <u>WORSHIP "LUCIFER"</u> <u>NOT THE ONE GOD</u>! ALL THEIR RITUALS ARE TO <u>PLEASE SATAN</u> AND YOU CAN SEE <u>THE SUN AT THE CENTRE</u> (TOP IMAGE) <u>AND SKULL WORSHIPPING CAN BE SEEN BELOW!!</u> IS <u>IT NOT SUN WORSHIPPING / PAGANISM / IDOLATRY?</u>

**ARE CRITICAL SCIENTIFIC CONCEPTS RELATED TO PAGANISM / SUN WORSHIPPING?**

**"SCIENCE IS RELIGIOUS" - SUN / gods WORSHIPPING!!**

❖ Science is assumed to be "<u>atheistic</u>" (nothing to do with any religion). It is believed to be the branch of knowledge based on observation and experimentation i.e., <u>proven facts!</u> BUT PLEASE WAIT A MINUTE......

❖ Why is it that the <u>Elements of Periodic Table</u> and <u>Planets</u> (wandering stars) has names <u>related to Greek / Egyptian gods? Is a coincidence by the scientific community?</u>

❖ <u>Are we not supposed to open our eyes and use our own "Intellect" to ask questions, as to why the scientific concepts are related to paganism BY ANY LOGICAL MEANS? Who were the people behind this? Where is it leading humankind? Does it make sense?</u>

❖ <u>Are we not believing in such a paganism, subliminally and unconsciously? If we accept everything fed in our mind from the mainstream, educational and media institutions!!!</u>

❖ Check this! <u>**HELIUM ELEMENT**</u> is named after "**HELIOS - SUN god**"

❖ **<u>THORIUM ELEMENT</u>** is named after "THOR - god of MYTHOLOGY". WHY??????

What is this going on!! <u>**Are we prepared to "wear the thinking caps" as a believer or remain contented with "What difference does it make for us, it is just Science???"**</u>

Are we ready to compromise and leave our generations to unconsciously believe in this paganism and not considering at all to be answerable before The One God on The Judgement Day? Please do contemplate as more evidence follows ahead.

# HELIOCENTRISM STARTED…….

COPERNICUS HELIOCENTRIC MODEL OF COSMOLOGY WAS PUBLISHED IN 1543 AFTER HIS DEATH

HE FACED RESISTANCE FROM THE RELIGIOUS COMMUNITY ABOUT HIS <u>PAGANISTIC DIVERGENT BELIEFS</u>! JUST CHECK THE IMAGE IN THE CENTRE, "<u>SUN ENGRAVED TOOL</u>" -WHY??? WAS HE JUST AN HONEST PHILOSOPHER IN SEARCH OF TRUTH OR HEAVILY INSPIRED BY <u>THE WRITINGS OF THE FIRST FREEMASON (ESSENTIALLY PAGAN) - PYTHAGORAS</u> (MENTIONED EARLIER) - ARE YOU GETTING THE LINK?

**POST "COPERNICUS" ERA MEANT TO DISGUISE OCCULTIC / PAGAN RELIGION AS "SCIENCE"**

**WHERE SUN IS THE CENTRE!!!**

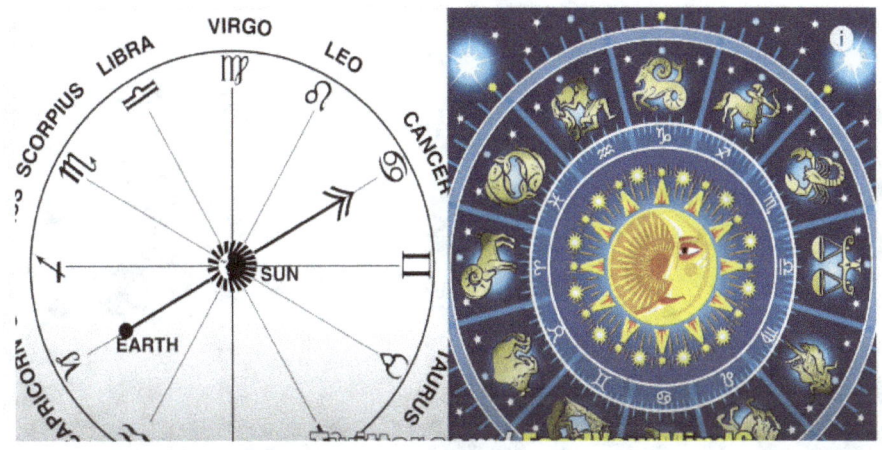

PSEUDOSCIENCE BEGINS TO BRAINWASH THE HUMANITY AND LEADING THEM TOWARD SUN WORSHIPPING / IDOLATRY GRADUALLY (OF COURSE UNCONSCIOUSLY) BECAUSE THEY ARE NOT SO STUPID TO DO IT BLATANTLY!!!

**SUN WORSHIP 2.0**

**COPERNICUS** - The infamous Freemason Catholic Priest led the humanity to believe in <u>spinning ball earth, heliocentric solar system model five hundred years ago</u>!! Was he serving the true teachings of the Bible or paganists i.e., sun worshippers?

Element "Copernicium" named after The Freemason Priest - Copernicus Why??? Was <u>he rewarded for diverting the mankind</u> from the true teachings of Holy Scriptures and Beliefs about the "Cosmology" designed By The Only One Lord God (Allah The Most High)!!

Before Copernicus the majority of mankind believed that the <u>Earth is flat and the sun and stars (planets) revolve around the Earth</u> and it is the "<u>center of universe" i.e. Geocentric Model</u>! This <u>has not only been accepted by the believers in Abrahamic Faiths / Scriptures</u>, rather In almost <u>20 Civilizations in human history</u>!

# ROMAN CATHOLIC CHURCH CONNECTED WITH THE BIG BANG THEORY!

**ROMAN CATHOLIC LUCIFERIAN (PAGANISTIC) CHURCH ENGAGED IN THIS BOGUS AND DEMONIC BIG BANG THEORY (NEVER PROVEN)!!**

Please see the connection that This Bogus Theory, never heard of is coming out of the Roman Church (essentially Pagan / Luciferian and nothing to do with Christian Religion, proving that everything came into being out of an explosion! Logically speaking any sort of explosion causes destruction "not Creation"! But propaganda through the powerful media and educational institutions at the highest level is so tremendous that the majority of the people are so brainwashed and indoctrinated that they fight to defend this Bogus

**Theory – Never Ever Proven!!** They even ignore the clear verses in The Holy Scriptures about The Creation Of Universe And Mankind, rather tend to interpret the holy scriptures form this "Bogus Pseudoscientific" (Fake Science) Concept!!!

**THEY, GRADUALLY FURTHER THEIR AGENDA FROM THE BOGUS BIG BANG THEORY (NEVER EVER A PROVEN FACT) TO ANOTHER DECEPTION -THEORY OF EVOLUTION!!**

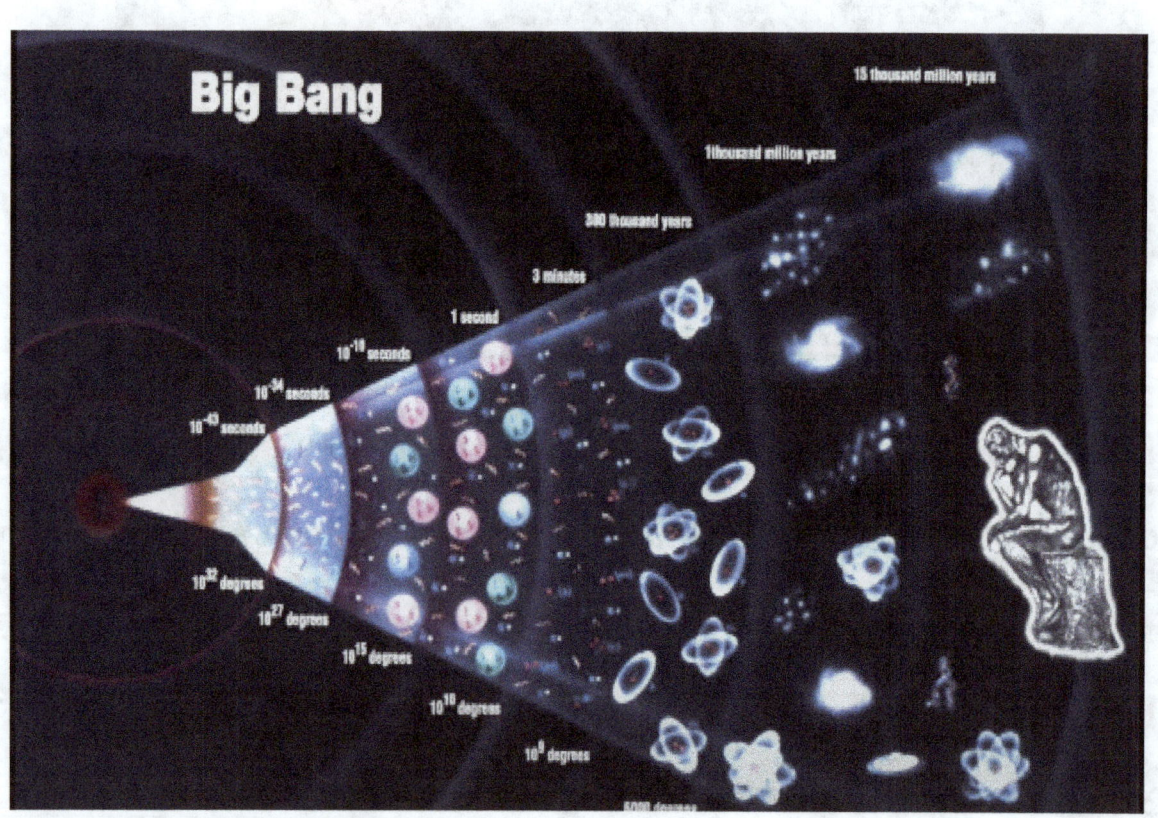

Just to prove that The Master Creation Of God [Allah The Almighty], "The Human Being" [along with The Glorious Earth and its resources meant for the human beings] is an "animal like non-purpose creature created out of series of accidents!" - courtesy, contribution of a pseudoscientist / pseudo philosopher - Charles Darwin, who put the nail in the coffin by writing "Origin Of Species"!!

This was the paradigm shift to astray mankind from THE PURPOSEFUL GOD's CREATION TO ACCIDENTAL MATERIALISTIC CREATION out of an explosion known as "monkeys"! - what utter nonsense and heresy!!

After five hundred years of brainwashing (starting from Copernicus) humanity got further astray from the true path and started believing in fake theories of outer space!! - media, comics and movies played a critical role for such a level of programming and indoctrination!!!

These Evil People wants us to believe that, We (Human Beings) [who actually are the master and purposeful <u>Creation Of God / Allah The Most High</u>] are only a "speck of irrelevant, non-purpose, materialistic dust In The Universe!

This gave rise to the <u>baseless theories of outer space having "Aliens"</u>! Please note that there is <u>No Space!</u> We live In "<u>A Closed Dome</u>", clearly mentioned in the scriptures and even understood by ancient civilizations thousands of years ago!!

There has been an established connection to Sun Worshipping since time and memorial from **Greek To Roman, Samarian And Egyptian Times** till we can see a clear connection with the "Evil Elite Pseudoscientific Community", who have been successful to "Spin" the majority that we are revolving and spinning around the Sun at a tremendous speed and it is the centre of our cosmology!!!

**IMPLEMENTING SUN WORSHIPPING THROUGH "HELIOCENTRIC**

**AGENDA" I.E. SPINNING BALL GLOBE EARTH AROUND THE SUN**

**AT THE SUBCONSCIOUS LEVEL!!!**

**SEE HOW THEY DERIVE HELIOCENTRIC FROM "HELIOS god" -**

**See the symbolism in the picture below and please ponder!!!**

**HELIOS SUN WORSHIP**
HIDDEN IN PLAIN SIGHT

IS THIS NONSENSE (BELOW), <u>SCIENCE OR PSEUDOSCIENCE</u> FOR PUSHING THEIR AGENDA BY <u>SUPPRESSING AN INQUISITIVE MIND?</u>

<u>THEY WANT THE MANKIND TO ACCEPT THIS AND GO AWAY FROM THEIR ONE ALMIGHTY CREATOR (THE TRUTH & REALITY) AND ADOPT PAGANISM / IDOLATRY TO LIVE A NON PURPOSE MATERIALISTIC ANIMAL LIFE!</u>

THE BRAINWASHING STARTS FROM KINDERGARTEN TILL GRAVE....

**EVENTUALLY LEADING MANKIND TOWARD** A DECEPTIVE, PURPOSELESS, MATERIALISTIC AND ULTIMATELY PAGANISTIC WAY OF LIFE, INITIALLY UNCONSCIOUSLY BUT EVENTUALLY HAPPILY BY CHOICE!!!

# "GEOCENTRIC MODEL" - A TRUTH WITH EARTH AS THE CENTRE OF OUR COSMOLOGY!

**PLEASE BE INFORMED THAT <u>ALL SEA AND AIR NAVIGATION IS DONE ON THE BASIS OF "TRUE FLAT EARTH MAPS" TILL DATE!!!!</u>**

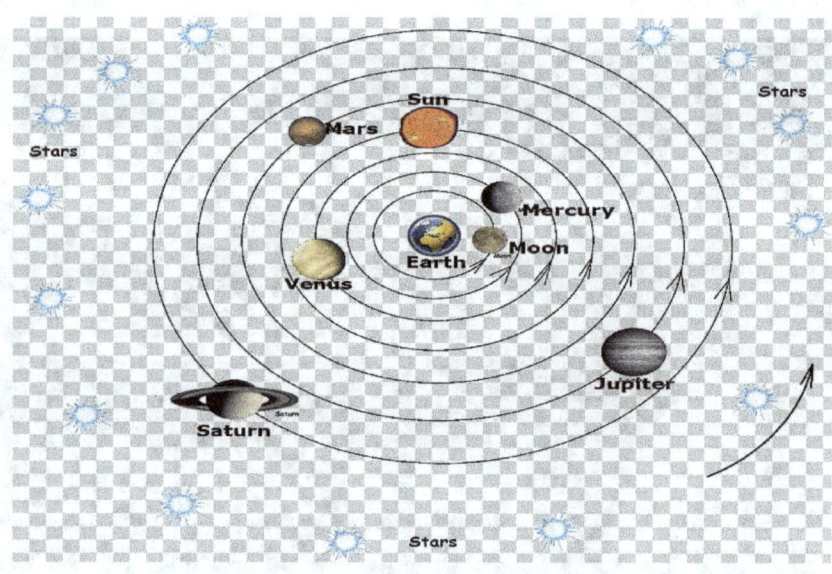

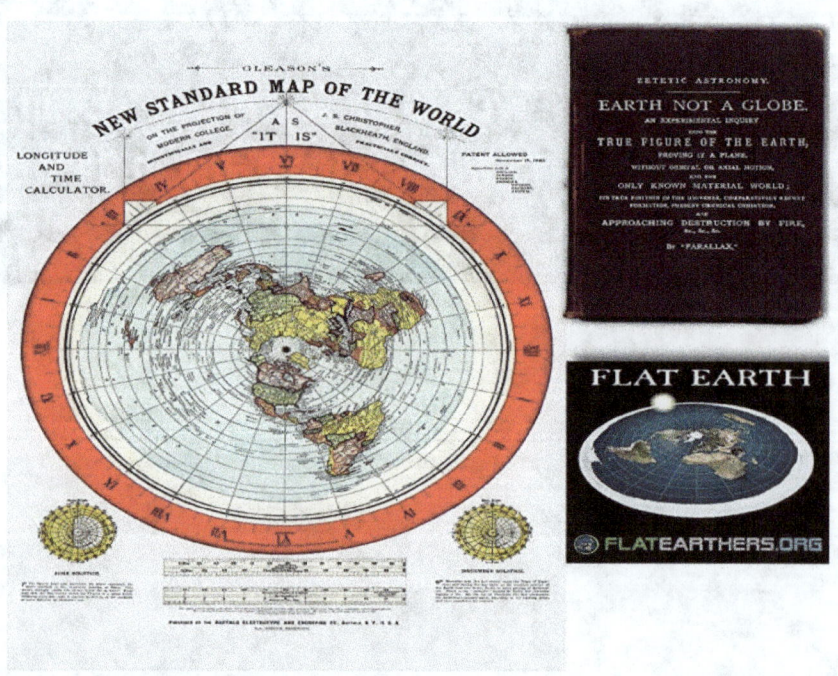

# Flat Earth Map - Gleason's 1892 New Standard Map Of The World

## 1943 Flat Earth World War II Map Polar Azimuthal Equidistant Projection

If you are <u>new and surprised to learn this concept</u> that the "<u>EARTH IS FLAT</u>" and really interested to explore more toward gaining knowledge i.e. <u>Truth about the Geocentric</u> (True Flat Earth) model then

please click the link below to read my blog:

FLAT EARTH vs GLOBE EARTH - IN SEARCH OF TRUTH FROM QURAN, BIBLE AND SCIENCE

But we will shed some light to explain about the "**Flat Earth Concept**" ahead**, as it is of critical importance and related to our subject!**

## SCIENCE VS SCIENTISM

**THIS LEAD TO THE DISASTER TO NATURAL QUERY OF AN INQUISITIVE MIND FOR EXPLORING THE REALITY ……..SCIENTISM!**

FB/DavidAvocadoWolfe

**SCIENTISM**
— An excessive deference to claims made by scientists or an uncritical eagerness to accept any result described as "scientific."
Also includes the concept of a SCIENTIFIC EGO
— only science has the right to describe the world around us.
The fanatical EGO-CENTRICITY of Scientism takes on many aspects of religious fundamentalism.

# SCIENCE
## VERSUS
# SCIENTISM

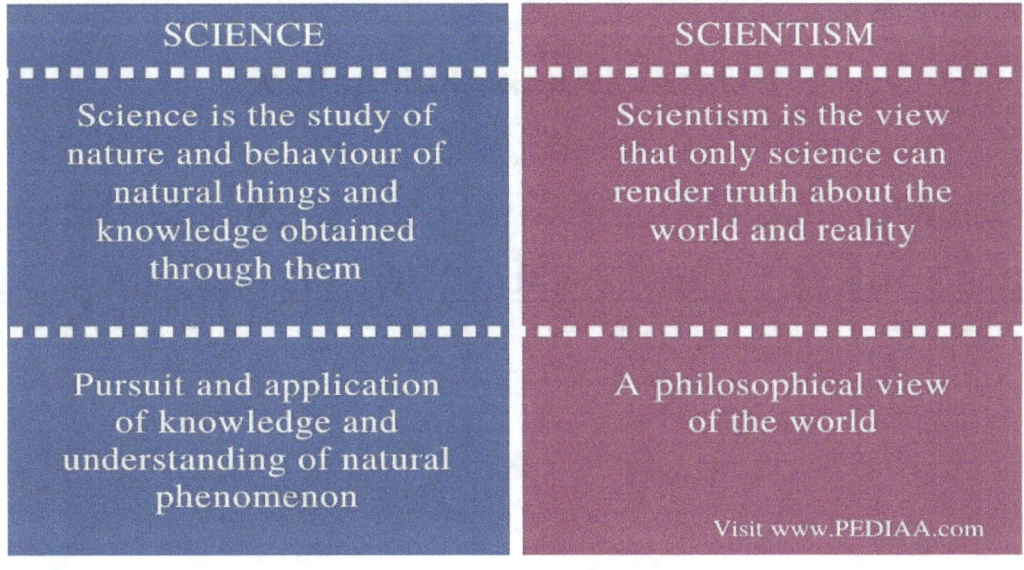

| SCIENCE | SCIENTISM |
|---|---|
| Science is the study of nature and behaviour of natural things and knowledge obtained through them | Scientism is the view that only science can render truth about the world and reality |
| Pursuit and application of knowledge and understanding of natural phenomenon | A philosophical view of the world |

Visit www.PEDIAA.com

If you are keen to learn more about the <u>Difference</u> between "Education" imparted by the mainstream institutions and "Knowledge" (which has unfortunately been evaporated from these institutions), and where the <u>current mainstream Education System is leading the mankind / coming generations</u> then,

you may like to please read my Book available at Amazon.com:

## "Importance of Acquiring Knowledge in the Modern Age"-

## Please click the link below:

Amazon.com: IMPORTANCE OF ACQUIRING KNOWLEDGE (علم) IN THE MODERN AGE: FROM HISTORICAL, CONTEMPORARY, AND ISLAMIC PERSPECTIVES eBook: Ala, Syed: Kindle Store

## CERN AND ITS SATANIC OBJECTIVES!

World's most advanced and complex Nuclear Research Facility on Quantum Physics - European Organization for Nuclear Research (CERN) involved in satanic objectives? Are they doing real science or trying to achieve some ulterior objectives??

"We are told by the (scientific community) that the purpose of <u>CERN</u> is to find the <u>Origins of Man and the Universe</u>. <u>They have stated that they want to open a doorway to another dimension and find a god.</u> Is it just a coincidence that CERN is short for (searching) the <u>horned god Cernuous?</u> Is it also a coincidence that <u>CERN has to go deep underground to do their "god" harnessing experiments? Cernunnos was the "god of the underworld".</u> There has been an ongoing conspiratorial dialogue about <u>what really is going on at CERN</u> and what the purpose of this collider really is. <u>Is it a high-powered dimensional tool to create star gates to</u>

**welcome entities (demons) into the world and could the attempts at doing such a thing destroy the earth?"**

→ Excerpts from an article: https://groundzeromedia.org/10-14-20-cern-and-the-gates-of-hell/

*"He [Stephen Hawking] knows the information that Tesla knew: they both know that there are entities that are providing information to people in this realm and that's where this advance technology is coming from, it is coming from the other side."*

*— Anthony Patch, interview extract.*

**MARK OF THE BEAST 666! CHECK THE LOGO OF CERN, ANY COINCIDENCE OR SHEAR DISPLAY OF DEMONISM / SATANISM BY THE SCIENTIFIC RESEARCH ORGANIZATION (SUPPOSED TO BE ATHEIST) FOR THOSE WHO CAN SEE!!!**

## LET'S TALK REALLY ABOUT "ROCKET SCIENCE"

ITS MASTER MINDS WERE TOTALLY OCCULTISTS, BLACK MAGICIANS AND OF COURSE PAGANS!!!

Aleister Crowley Jack W. Parsons Ron L. Hubbard Werhner von Braun  Walt Disney
What does an avowed Occultist, the founder of JPL and NASA, a Science-Fiction Writer
and founder of the Church of Scientology; an Ex-Nazi turned America's foremost
Rocket Scientist and Walt Disney, founder of the Magic Kingdom - have in common?
oh, you might be surprised the extent of their collaborative efforts in Babalon (sic -
'Babylon') on behalf of Horus, Mars and alien life forms...yes, you might be real
surprised - indeed, this is 'stranger than fiction' - then again, magic, aliens, space
travel, and 'science' have a lot more in common than you might think - then again, do
Americans do all that much thinking?  Yes, you'll be very much surprised.

Just do some research about the profile of <u>these occultists</u>, most wicked pagans and see <u>how majority believe in and cherish them about their so-called scientific contributions writings and philosophies, without even thinking</u> that where were these evil and deceptive people leading the mankind? MAGIC, ALIENS AND PAGANSIM!!

<u>E.g., Jack Parsons</u> (Rocket Scientist) **made this fact even clearer when he started to develop a growing interest in magic and the**

**supernatural. By the late 1930s, he had begun frequenting nightly meetings** of the Ordo Templi Orientis, **an occult society** that met in nearby Los Angeles. The OTO, as it is known, was created by the English occultist **Aleister Crowley, a heroin-addicted, sexually adventuresome, God-profaning master of the dark arts**, who the tabloids had christened "The Wickedest Man in the World."-

Source:https://www.vice.com/en/article/vvbxgm/the-last-of-the-magicians

Reference:

**Dajjal, Technology, Magic, Demonic Culture**

**Please watch: https://youtu.be/XuEBMQZ4Vd4**

**NASA - A GREATEST PSYOP IN HUMAN HISTORY!!!**

→ Established in 1958 **TO BE THE GATEKEEPERS OF KNOWLEDGE (TRUTH),** after **the discovery of huge land mass, the size of USA** by **Admiral Byrd in 1946**, **tend to bust the Ball Earth fake Heliocentric solar system theory** **(if allowed to explore)**, **followed by** the **Treaty of Antarctica** signed by **12 nations in 1959 meant to FORCEFULLY CONTROL / RESTRICT unbiased research** (real science) **beyond 60° latitude - WHY??**

→ It is also meant to **control the knowledge** about **True Cosmology** and **restrict humankind to gain access about it!** Hence, **NASA** has propagated whatever (fiction till date) must be accepted without a question!! (For more knowledge, please refer to my blog on this subject, mentioned in the slide earlier)

→ **It has a budget of around USD 16 billion** (taxpayers money) and **if they are exposed of their fake "Spinning Ball Earth" propaganda, fake Moon Landing, fake Mars missions etc. then they will fall "FLAT" and don't justify their existence for a minute to consume billions of dollars by showing us cartoons! Do you**

**think these so-called EVIL FREEMASONS, WHO WANT TO CONTROL THE RESOURCES AND HAD BEEN PLANNING AND BRAINWASHING THE MANKIND FOR CENTURIES WILL LOOSE THEIR CONTROL??**

→ This presentation is meant to **expose only the linkage of the NASA Astronauts, key personnel with the Freemasonry / Paganism / Sun Worshipping!!**

## NASA's CONNECTION WITH FREEMASONRY [OF COURSE PAGANISM!]

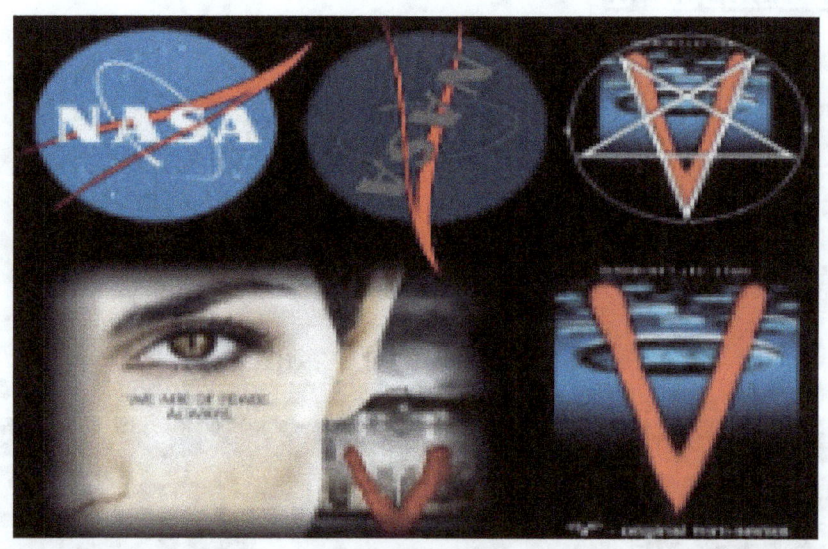

NASA's logo is a giant red forked serpent's tongue overlaying the starry heavens. Serpents, and specifically their forked tongues, have long been associated with lying, deceit, cleverness, two-facedness, manipulation, and with Satan, the Devil. Why would the National Aeronautics and Space Administration choose this as their official logo?

In <u>1969 the Sun-worshippers</u>, aboard a craft named "<u>Apollo</u>" after the Greek Sun god, claimed to land on, and <u>thereby spiritually and physically "conquer," the Moon</u>.

The truth is that these so-called "astronauts" <u>never landed on moon,</u> rather it was a <u>MGM Studio "psyop production in Navada Desert"!!</u>

# GLOBE LIES

**LIE**

## THEY ARE IN SPACE - ON OTHER PLANETS
## DISCOVERING PLANETS AND GALAXIES
## SPENDING OUR MONEY - DOING TONS OF STUFF

- NOT TAKING PICS OF EARTH
- NOT PROVING THE MOON LANDINGS
- NOT SHOWING ECLIPSES FROM SPACE
- NOT GIVING US 24 HOUR WEB CAMS FROM SPACE
- NOT GIVING US WEBCAMS FROM THE MOON

- NOT ALLOWING PRIVATE CORPS UP
- NOT GIVING US TRUE INFORMATION
- NOT GIVING US TRUE IMAGES
- NOT GIVING US DISCOVERIES OR INVENTIONS
- NOT HELPING HUMANITY

34:51 / 1:08:50

## THE GLOBE LIE

- YOU ARE SPINNING 1000 MPH
- YOU ARE FLYING THROUGH SPACE 66K MPH
- YOU ARE UPSIDE DOWN AND RIGHTSIDE UP
- YOU ARE IN AN ENDLESS UNIVERSE
- THE SUN IS 92 MMA MILES AWAY AND IS NUCLEAR
- THE MOON IS 250K MILES AND CAN BE LANDED ON
- THE ATMOSPHERE IS STUCK TO EARTH AS IT SPINS
- THERE ARE BILLIONS OF EARTH LIKE PLANETS
- THE SUN IS THE SAME AS THE STARS
- STARS ARE TRILLIONS TO QUINTILLIONS OF M.A.
- WE COULD BE STRUCK ANY MOMENT BY A METEOR
- YOU ARE ACCIDENTAL AND A PRODUCT OF STARDUS
- YOU ARE A PRODUCT OF EVOLUTION

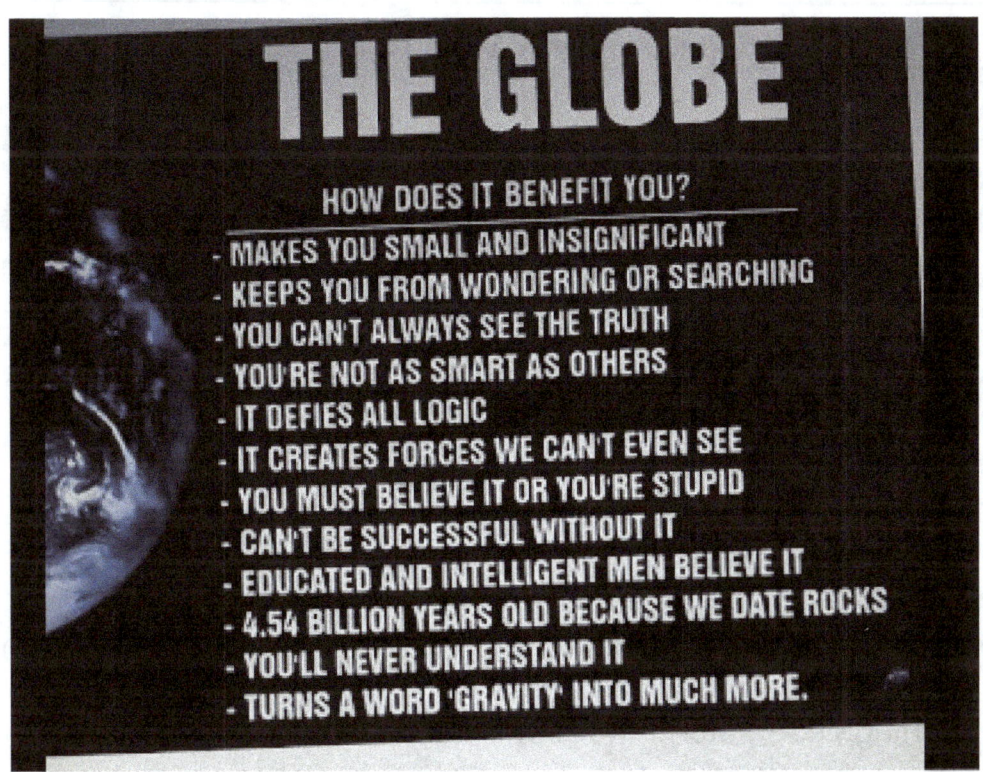

## IMPORTANCE OF LEARNING ABOUT FLAT EARTH (TRUTH) - GEOCENTRIC MODEL

Now, I feel that it is an obligation upon me to explain this subject, especially for those who are not aware of it. I had been studying this subject of Earth being "**Flat**" or "**Globe**" for **couple of years**. One might wonder whether this topic is worth studying or discussing, as almost everybody (except minute minority i.e. "**Flat Earth propagators**") from all walks of life confirm to the level of their belief and study that "**Earth is an oblate / spheroid shape planet spinning around the sun**". But is it true?

From my perspective, it is very important to explore (to the best of my efforts) about the **true cosmology** of the Universe and especially about our wonderful "**Earth**", as our The One Lord God (Allah The Most High) invites His slaves (entire mankind) to do so in the following verses of Holy Quran:

"And to Allah belongs the dominion of the heavens and the earth, and Allah is over all things competent. Indeed, in the creation of the heavens and the earth and the alternation of the night and the day are signs for **those of understanding (critical thinkers)**. Who remember Allah while standing or sitting or [lying] on their sides and give thought to the creation of the heavens and the earth, [saying], "Our Lord, **You did not create this aimlessly**; exalted are You [above such a thing]; then protect us from the punishment of the Fire." – [Chapter 3, Verses 189-191]

I also pray to Allah The Almighty: **O Allah! show us the truth as true and inspire us to follow it. Show us falsehood as falsehood and inspire us to abstain from it - Ameen!**

## PERCEPTION DECEPTION:

There is a famous saying that "**many people look (*appearance\**) but very few can see** (*reality* *)". We have been lied about many things by our governments (which itself means "**ment** =mind + **govern** = control") and all sorts of mainstreams public / private institutions (education and media) in order **to brainwash and indoctrinate us** (majority), so that we **may not be able to think critically and raise questions / challenge** about anything from our cradle / kindergarten to grave, rather simply believe in everything from the authorities and submit to them as their slaves!

(*) writers' comment / insertion

**Note:** Please remember that nothing happens by accident rather purposely done to enslave the masses! Once Franklin Delano Roosevelt (US President) said: **"In politics nothing happens by accident. If it happens, you can bet it is was planned that way"**

**My humble research led to following conclusions (a few from a long list):**

- **Earth is Flat** - (Geocentric model is a Truth!)

- **Globe Earth - (Heliocentric model,** which is named after **Greek god "Helios",** hence the proponents of this theory **are essentially pagans,** who have taken majority of the human being on a dangerous ride!**)**

- **Moon landing is crap (a fiction movie!)**

- **Big Bang is just a baseless theory and there is no SPACE!**

- **Gravity is a False / Fake concept!**

- **NASA is a Masonic Psyop (the biggest cover up in human history)!**

- **Evolution Theory is an utter garbage!**

- **Einstein's Relativity Theory was a mathematical garbage** (stated by Nikola Tesla), while Nikola Tesla was a God fearing, genius and a true scientist, but majority do not know and intend to learn about him and his inventions!

There are more than 200 proves that are related to Flat Earth and its related concepts, which challenges and discard the fake & fraudulent concepts of Globe Earth Model!!

Reference: A book "200 Proves that Earth is Flat" by Eric Dubay.

## SERIOUS CONSEQUENCES!!!

→ Had there been no "Globe Earth" fake concept, there would have been no Big Bang (theory)

→ No Big Bang, then there would have been no Evolution (theory)

→ No Evolution, then masses would have believed in an intelligence design of our One Creator (Allah The Al Mighty) and we would have truly known our great purpose of our existence on Earth and importance of Earth in the Universe, as a focal point in His Master Plan!

→ This ultimately would not have led to the current situation of materialistic, godless, and aimless way of life (worse than animals), just serving the authorities and have no or least idea about our life here-after and The Day of Judgement!!

# EVIDENCE FROM HOLY SCRIPTURES ABOUT FLAT EARTH AND RELATED CONCEPTS

## Quranic Verses Proving Flat Earth:

Please find below the references for your kind understanding:

[51:48]

"And the earth have We **spread out wide** - and how well have We ordered it!" -

## Some More Verses About our Nearest Sky:

[20:53]

"He Who has, made for you the earth like a **carpet spread out (flat)**; has enabled you to go about therein by roads (and channels); and has sent down water from the sky." With it have We produced diverse pairs of plants each separate from the others."

[79:27 -30]

"Are you a more difficult creation or is the heaven? Allah constructed it. He raised its **ceiling (dome)** and proportioned it. And He darkened its night and extracted its brightness. And after that **He spread (flattened) the**

**earth."**


[21:32]

"and [that] We have set up the **sky as a canopy well-secured (dome)**?
And yet, they stubbornly turn away from [all] the signs of this [creation]!"


[78: 12 -19]

"and built above you **seven strong firmaments (domes)**, and placed

therein a hot, shining lamp, so that We may thereby bring forth grain and

vegetation, and gardens dense with foliage? Surely the Day of Judgement

has an appointed time; the Day when the Trumpet shall be blown, and you

will come forth in multitudes; and when the sky shall be opened up and will

become all doors;"

* Dome  (Heaven / Samaa) = mentioned by me


**Quranic Verses Proving the Sun and Moon are revolving around our**

**Flat Earth (Earth is Stationary / Fixed Platform):**


Please find below some of the references for getting the Truth from The

Holy Quran:

**[14:33]**

Who subjected for you the **sun and the moon and both of them are constant on their courses,** Who subjected for you the night and the day [18:17]

Had you seen them in the Cave it would have appeared to you that **when the sun rose, it moved away from their Cave to the right; and when it set, it turned away from them to the left,** while they remained in a spacious hollow in the Cave. This is one of the Signs of Allah. Whomsoever Allah guides, he alone is led aright; and whomsoever Allah lets go astray, you will find for him no guardian to direct him.

**[18:86]**

"until when he reached the very limits where **the sun sets**, he saw it setting in dark turbid waters; and nearby he met a people. We said: "O Dhu al-Qarnayn, you have the power to punish or to treat them with kindness."

**[18:90]**

"until he reached the limit **where the sun rises,** and he found it rising on a people whom We had provided no shelter from it."

**[20:130]**

"So bear patiently with what they say. Glorify your Lord, **praising Him before sunrise and before sunset**, and in the watches of the night, and glorify Him and at the ends of the day that you may attain to happiness."

**[21:33]**

"It is He Who created the night and the day, and **the sun and the moon. Each of them is floating in its orbit.**

**[22:18]**

"Have you not seen that all those who are in the heavens and **all those who are in the earth prostrate themselves before Allah; and so do the sun and the moon, and the stars** and the mountains, and the trees, and the beasts, and so do many human beings, and even many of those who are condemned to chastisement? And he whom Allah humiliates; none can give him honour. Allah does whatever He wills."

**[25:45]**

"Have you not seen how your Lord lengthens out the shadow? Had He willed, He would have made it constant, but **We have made the sun its pilot"**

**[31:29]**

"Do you not see that Allah makes the night phase into the day and makes the day phase into the night and has subjected the sun and the moon to

His will so that each of them is pursuing its course till an appointed time? (Do you not know that) Allah is well aware of all that you do?"

[35:13]

"He causes the night to phase into the day and the day into the night, and **He has subjected the sun and the moon, each running its course to an appointed term.** That is Allah, your Lord; to Him belongs the Kingdom; but those whom you call upon, apart from Allah, possess not so much as the skin of a date-stone."


[36:40]

**"Neither does it lie in the sun's power to overtake the moon nor can the night outstrip the day. All glide along, each in its own orbit."**


[39:5]

"He created the heavens and the earth with Truth, and He folds up the day over the night and folds up the night over the day. **He has subjected the sun and the moon, each is running its course until an appointed time.** Lo, He is the Most Mighty, the Most Forgiving."

**[50:39]**

"Hence bear with patience whatever they say, and celebrate your Lord's glory **before the rising of the sun and before its setting;"**

**[55:5]**

**"[At His behest] the sun and the moon run their appointed courses;"**

**[91:2]**

**"And by oath of the moon when it follows the sun"**

**Biblical Verses Proving Flat Earth Concepts (Earth is Stationary / Fixed Platform):**

We are presenting some of the verses out of many verses, just to give the readers an idea about the Truth from the Holy Bible:

**Book of Genesis 1 - Verses: 1** In the beginning God created the heaven and the earth. 2 And the earth was without form, and void; and darkness was upon the face of the deep. And the Spirit of God moved upon the face of the waters. 3 And God said, Let there be light: and there was light. 4 And God saw the light, that it was good: and God divided the light from the darkness. 5 And God called the light Day, and the darkness he called

Night. And the evening and the morning were the first day. 6 And God

said, Let there be a firmament (dome / Sky) in the midst of the waters, and

let it divide the waters from the waters.

**Genesis 16-18** God made two great lights—the greater light to govern the

day (moon) and the lesser light to govern the night (moon). He also made

the stars. 17 God set them in the vault of the sky to give light on the

earth, 18 to govern the day and the night, and to separate light from

darkness. And God saw that it was good.

**Genesis 19: 23** By the time Lot reached Zoar, **the sun had risen over the**

**land**

**Psalm 148:3** Praise ye Him, **sun and moon: praise Him, all ye stars of**

**light**

**Ecclesiastes    1:5** The sun   also ariseth,    and    the    sun goeth down,

and hasteth to his place where he arose.

**1 Chronicles 16:30** Tremble before him, all the Earth! The world is firmly established; it cannot be moved.

**Psalm 93:1** The Lord reigns, He robbed in Majesty, the Lord is robed in Majesty and armed with strength; indeed, the world is established, firm and secure.

**Exodus 22:3** but if it happens after sunrise, the defender is guilt of bloodshed

**Zechariah 1**

11 And they reported to the angel of the Lord who was standing among the myrtle trees, "**We have gone throughout the earth and found the whole world at rest and in peace.**"

## MIND BLOWING SCIENTIFIC REFERENCES FOR EXPOSING THE TRUTH:

## Mainstream Media Finally Admit Flat Earth!

Please click to watch (few minutes) and seek the truth - https://youtu.be/KeWQ8hWl7DM

## Our Flat Earth Has No Curvature! - By Eric Dubay

Please click to watch (few minutes) and seek the truth - https://youtu.be/DPGW9blx_e8

## Globe Earth LOSES In COURT vs Flat Earth!! (Judge's Decision!!)

Please click to watch (few minutes) and seek the truth -

https://youtu.be/DZDZVZHa-cg

**Court Proceedings:** Please read –

https://www.facebook.com/notes/377143260124586/

**Buzz Aldrin asked: Did We Go To the Moon? or is the Earth Flat with A Dome?**

Please click to watch the interview (few minutes) and telling the truth - https://youtu.be/xt9RfXCKYG4

**SECRET DISCOVERY in Antarctica HIDDEN in PLAIN SIGHT!!**

Please click to watch (few minutes) and seek the truth - https://youtu.be/3laF7GeeRAQ

**I WOULD HUMBLY INVITE THE RESPECTED READER TO DO SOME RESEARCH INDEPENDENTLY TO CHECK IF I AM RIGHT!**

**AS IGNORANCE AND BLIND FOLLOWING IS NO EXCUSE IN THIS AGE OF DECEPTION!**

## NASA's CONNECTION WITH FREEMASONRY [OF COURSE PAGANISM!]

An inordinate number of <u>NASA astronauts, the current propagators of the globalist" heliocentric doctrine", are/were admitted Freemasons as well</u>. **John Glenn, two-time US senator and one of NASA's first astronauts** is a known Mason. **Buzz Aldrin Jr.**, the second man <u>to lie about walking on the moon is an admitted, ring-wearing, hand-sign flashing 33rd degree Mason from Montclair Lodge No. 144 in New Jersey</u>

*C. Fred Klein Knecht, <u>head of NASA at the time of the Apollo Space Program, is now the Sovereign Grand Commander of the Council of the 33rd Degree of the Ancient and Accepted Scottish Rite of Freemasonry of the Southern Jurisdiction.</u>*

The United Nations, the New World Order government headquarters, built on land donated by 33rd degree Freemason John D. Rockefeller,

is represented by a **logo/flag** which clearly depicts **a Flat Earth divided into 33 sections!** **There are 33 official degrees of Scottish-Rite Freemasonry, and the UN flag features a Flat-Earth divided into exactly 33 sections!**

## KEY QUESTIONS?

➢ Why would the United Nations founders choose a logo/flag of a Flat-Earth map divided into 33 sections?

➢ How is it that C. Fred Klein Knecht, the head of NASA, retired and immediately became the head of the 33rd degree of Freemasonry?

➢ How is it that all the ancestors of the ball-Earth theory and so many NASA astronauts are all Freemasons!?

Can you appreciate the connection of NASA with paganism plain as daylight?

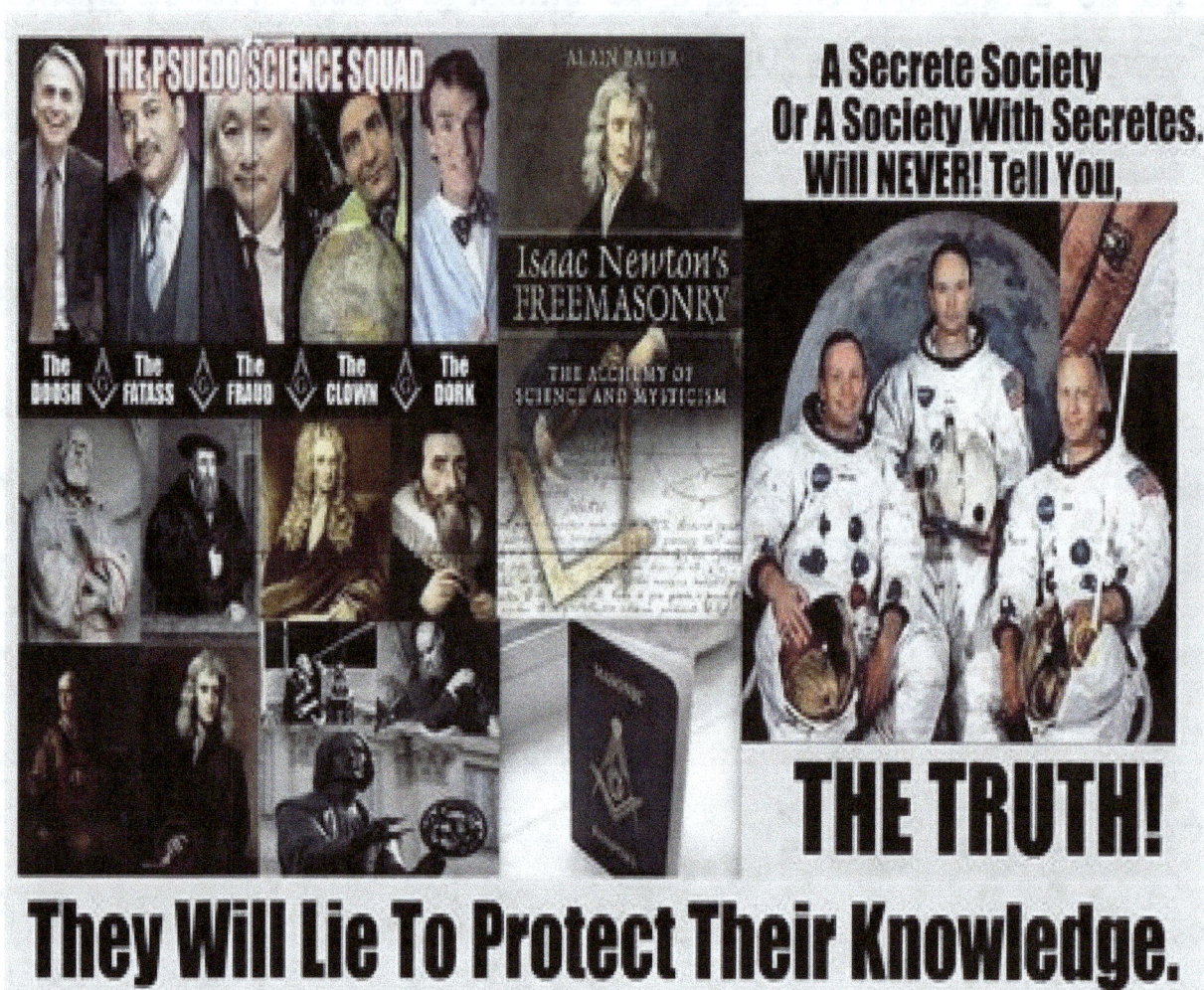

# ISLAMIC PERSPECTIVE

شَيْءٍ كُلِّ عَلَىٰ وَهُوَ الْمُلْكُ بِيَدِهِ الَّذِي تَبَارَكَ
أَيُّكُمْ لِيَبْلُوَكُمْ وَالْحَيَاةَ الْمَوْتَ خَلَقَ الَّذِي ) ١ ( قَدِيرٌ
الْغَفُورُ الْعَزِيزُ وَهُوَ عَمَلًا أَحْسَنُ

Blessed is He in Whose hand is the Sovereignty, and He is Able to do all things. Who has created death and life that He may test you which of you is best in deed. And He is the Almighty, the Oft-Forgiving.

Al Quran - Chapter Al Mulk: Verses 1-2

Our current Age, which is at the peak of Trials and Tribulations (End Times) is to test mankind, so that only those who have **"Faith" in One Lord God / Allah SWT** (with deep sincerity in heart), **His Final Messenger Muhammad** (peace be upon him and his progeny) and **completely follow the Truth** (not partially or in a secular manner) **may survive the test and enter in the Heaven /Jannah!**

All those who astray and follow paganism (of any form, consciously or do not use their intellect to understand its nature and threats, despite warnings) will all end up in the single (worldwide) melting pot of godlessness and ultimate may face severe consequences on the Judgement Day!

## Allah The Most High warns in the Holy Quran:

**"Many are the jinns and men we have made for Hell: They have hearts wherewith they understand not, eyes wherewith they see not, and ears wherewith they hear not. They are like cattle,- nay but they are worse! for they are heedless (of warning)"** -

**Surah Al A'raf- Chapter 7:179**

## SHIRK / BLASPHEMY - THE GREATEST AND UNFORGIVABLE SIN!!

* In **Islamic and other Abrahamic religions,** the biggest "Sin / Gunah e Kabira" is the "Shirk / Blasphemy / Associating partners in One Deity", which Allah SWT (One Lord God) will never forgive (**unless sincere repentance is sought before the moment of death**).

* He SWT says in the Quran: " **Verily, Allah forgives not (the sin of) setting up partners (in worship) with Him, but He forgives whom He wills, sins other than that, and whoever sets up partners in worship with Allah, has indeed strayed far away**" - Chapter 4, Verse 116)

* There are various forms of "Shirk / Blasphemy," classified under **"Shirk e Jali (Open)" and "Shirk e Khafi (Obscure /**

**Hidden)**". The **Hidden form of Shirk is very difficult to understand** as mentioned by our beloved Prophet Muhammad (peace be upon him and his progeny): "The ordinary polytheism (hidden shirk) is more hidden among this nation than the track of a black ant over a black stone on a dark night"

– Reference: Musnad Ahmad

In case you wish to learn in depth about **Shirk / Blasphemy** and its forms, please click the link below to view my detailed presentation on the topic: -

https://realityandtypesofshirk.blogspot.com/2016/08/dear-brothers-in-islam-please-find-link.html

## WE ARE THE SLAVES OF ONE LORD GOD (ALLAH THE MOST HIGH)

* Allah SWT (One Lord God) is the **only One Who** has the "<u>right to be worshipped / Ibaadah alone in totality</u>". <u>NOTHING SHOULD LEVEL OR COME ANYWAY NEAR TO THIS EXCLUSIVE RIGHT OF SUBMISSION BEFORE ONE LORD GOD / ALLAH SWT!</u>

* It is very important to understand that wherever Al Quran refers to word "**Abd (عبد)**", it must be understood as "**Slave**", then can we really appreciate the true understanding of "**worship / ibadat**" i.e. <u>**total submission and obedience to Allah SWT in every spheres of life!**</u>

[TOTAL SUBMISSION AND OBEDIENCE TO ONE LORD GOD (ALLAH SWT), NO CHOICE WHATSOEVER!]

\* We need to have conviction that One Lord God / Allah The Most High **is The Supreme Authority of Allah SWT, because He is The Creator, The Great without any Comparison, The Sovereign and The Law Giver!!**

## CONCLUSION

A PERSON (BELIEVER OR A NON-BELIEVER) CAN BE A "SUN WORSHIPPER / PAGAN" ……...

→ **IF** - he / she <u>considers the Freemasons</u> as mere society **simply doing some sort of philanthropic work** for the betterment of the society.

→ **IF** - he / she believes that the "<u>Earth is a Globe Spinning</u> at 1100 miles per hour on its axis and 66,000 miles per hour around the Sun"!!**

→ **IF** - he / she believes in the fake and never proven "**The Big Bang Theory**"

→ **IF** - he / she believes in the bogus and fraudulent "**Theory of Evolution**"

72

→ **IF** - he / she believes and blindly accepts whatever is propagated / disseminated by NASA is truth AND **must be accepted without question!**

→ **IF** - he / she believes and blindly follows whatever "mainstream science" says is right without questioning!

Please note that I am not at all against "Real Science / Enquiry" i.e. TRUTH and Research for the betterment of humanity, but if it is coming from the pipe-smoking Freemason scientist / philosophers and their allies / institutions obsessed to deceive and control the humankind, then there is a HUGE problem and "I will resist" by attempting to make the humanity aware of the deception   and eventually achieve their "mental freedom" to the best of my abilities!!

Now, would you agree or disagree with my million-dollar question? Kindly no need to tell me, just please ponder, open your eyes, keep exploring and come to The TRUTH sent by our One Lord God (Allah the Most High) and Creator and save yourself and your generations from eternal punishment!!

Thank you for your kind attention, patience,

and support!


With best regards,


Jazak Allah khairan kaseeran!


For your kind feedback


ahsenala@gmail.com

www.ingramcontent.com/pod-product-compliance
Lightning Source LLC
Chambersburg PA
CBHW081534220526
45467CB00010B/3187